On The Right Track

A Student's Memoir of Research,
Advancement, and Holding on to Hope

JENNIFER LEE SCHWARTZ

On The Right Track,
A Student's Memoir of Research, Advancement, and Holding on to Hope

Jennifer Lee Schwartz

Copy Editor: Mary Perrotta Rich
Cover Design: Jeanine Henning

Follow Jennifer at www.jenniferleeschwartz.wordpress.com

ISBN-13: 978-1511472395
ISBN-10: 1511472391

Library of Congress Control Number: 2015904975
CreateSpace Independent Publishing Platform
North Charleston, South Carolina

For my mother and for Karolena,
May good health find you always

ACKNOWLEDGMENTS

To my Science Research teachers: Mr. Yagid, Mr. Hughes, and Mr. Gleason: thank you for making Science Research such a wonderful part of my life. I would like to thank the teacher who guided me through the journey of self-publishing and becoming an author in general: Mr. Holmes. To my fellow students who worked on projects as well: Sarah, John, Alexandra and Siqiao, I wish you the best of luck with your future endeavors. Thank you for the edits and help along the way.

I would like to thank all of my friends in the Science Research program: the juniors and seniors who helped guide me through every step of the process and my fellow sophomores who made it so enjoyable. Thank you: Sarah, John, Erin, Nick, Esther, Megan, Mike, Isabelle, Natasha, Will, Kat, Carla, Kelly, Richie, Ian, Hana, Liam, Randy, Cal, Aidan, Carey, Izy, Lily, Kristian, Adam, Tim and Chris. To all of the researchers who have helped me with my understanding of the disease and modern treatments, whether or not they are aware of the impact they have had on me, thank you: Dr. Green, Dr. Wang, Dr. Opferman, Dr. Letai, Dr. Glaser, Dr. Pandey, Dr. Martinou, Dr. Certo and especially Dr. Alan Eastman for all of the help he has given me.

Also, to the teachers and friends who helped this book progress to this final copy, especially Mrs. Phillips, Mrs. Donnelly, Ms. Waltuch, and Mr. Spiegelman, thank you for taking time for me. Thanks to Ms. Terhaar for making science fun for so many at a young age. Thanks to Mary Perrotta Rich for spending so much time helping me with this project.

I would like to thank Kelly, Karolena and their family for telling me their stories and allowing me to share them with you.

My family has offered their strongest support. Even if they do not completely understand the biological processes, they are here for me every part of the way. Thank you Mom, Dad, Jackie, and Chelsea.

Best wishes, Jen

"They always say time changes things, but you actually have to change them yourself."

Andy Warhol

CONTENTS

1

CANCER

An introduction to my life

A scientist might say that cancer is an abnormal growth of cells which evades death, allowing it to spread. However, a patient would describe it as a devastating illness that devours the body and causes excruciating pain. Neither of these definitions characterizes cancer completely. As a child, all I knew about cancer was that it was hurting my mom. I was five years old when my mom was diagnosed with Stage Four Hodgkin's lymphoma. I did not know what cancer meant and had no idea about the kind of uncomfortable treatments my mother would have to go through. I watched her hair fall out and I gave her massages to make her feel better. Being so young, her suffering did not register with me, but the pain she felt could be seen in her tired eyes reading me to sleep at night. Mommy was not okay.

Cancer patients are the only people who can ever really understand what it is like to have this foreign growth taking over their bodies. At times it seems as though the public's understanding of cancer is as misconstrued as mine was as a child, despite efforts at support and awareness.

There are so many different types of cancer, but most people are only aware of the few kinds that have affected their friends and family. Treatments and diagnoses vary so greatly, that no one story can capture the complete essence of cancer. Cancer is common--most people know someone who has been affected by it, and the story that it entails. There are stories of triumph, sadness, laughter, devotion, love, and devastation.

Even movies tell stories of cancer in different lights: battles of needles and chemical inhibitors exhibit the need to fight through the treatments. Support groups may even lead to a romantic plot twist for films. Movies do not always show the pain: the endless days cooped up in bed, the vomit, the blood or even the needles that inject the treatment. When the disease is depicted in film, some aspects of it are captured, but most of the details are not seen. Facts often go untold, and even the most informative films often glaze over the more traumatic parts. The scientific side of cancer is not necessarily box-office worthy.

Today, I know cancer is a seemingly endless struggle in which people undergo significant torture to fight to reach remission. Their cure may not be the end of their journey, but the end of their current suffering.

As a high school student, the questions flooding my mind are usually about my future. I worry about finding success through high-caliber schoolwork, near perfect SATs, AP classes, and attending a competitive college. To me, however, success is increasing the well-being of the world and the lives of the people who inhabit it. Visions of blue scrubs and white sneakers, yellow face masks and purple gloves, a patient open on the table with the exposed tumor a sickly pearl color, fuel my work ethic. Science is what allows people to innovate. Every piece of progress in cancer research allows these patients to live happier lives. This book is a reflection on my first steps toward lighting a match to ignite hope.

2

CURIOSITY

The gateway to research

A couple years after my mom went into remission, I knew I was going to attend our town's annual walk-a-thon for cancer. I saw a poster for it and thought I would be supporting a great cause with my friends. Together we could help people. I was young, and I knew that I was helping by just showing up. I did not want others to suffer like my mom had, so I walked laps "for cancer." Vast seas of people gallivanting to the finishing lap made me proud of my town for making a difference in the world: I was surrounded by people who knew about cancer and wanted to do what they could to support the cause. Speeches and ceremonies commenced, making many adults cry.

After attending this, I felt like I had made a difference, so my yearly attendance prevailed without any research into the organization or where the money we raised would go. While these organizations do donate some money to cancer research and treatments, to many patients money is of the least importance.

Patients in a stable economic state often need more than money. Hope inspires, whereas money only catalyzes physical progress. Anonymous donations from strangers who celebrate life with little recognition of the cancer patients are not always life changing.

I was in the middle of Mr. Gleason's freshman Biology class, where my friend Kathryn and I were planning the details of our walk-a-thon team that year. Kathryn and I had been working out the logistics of the evening: Allison was on snacks, Kate was on tent duty. Drowning out the pretzels or cookies debate, I turned my focus back to Biology. Another teacher had just entered and stood in the doorway. He was tall to say the least; his body practically filling the doorway in an intimidating manner.

"Class, this is Mr. Yagid, he is one of the other Science Research teachers," Mr. Gleason said before he pulled Mr. Yagid aside to discuss how a student of Mr. Yagid's was going to travel to Sweden for her research on obesity. At first, I was confused. There already is a cure to obesity, I thought: diet and exercise! Apparently there was a lot that I did not know about obesity and even about how research worked past the middle school experiment level. Natasha had been trying to convert "bad fat" to "good fat" with a drug that would increase calorie burning in mice. The creativity of her ideas made me wish people like her were working in fields like cancer research as well, where treatment was less known.

Questions swarmed in my head; *maybe cancer treatments like this are being worked on, I have no idea*. Shaking off the idea that I was ignorant about cancer research, I reassured myself; *my mom had it, of course I know about the disease.* Did I really comprehend it though? The complexities of the disease were still quite unclear to me. The basis of my cancer knowledge thus far was only all that I had learned in Biology this year. Freshman Biology Honors courses teach about a large range of topics in a short period of time. It touches on various concepts and defines vocabulary.

> **Cancer**, *noun*: when cells mutate and divide without regulation from the cell cycle

Biology class gives students necessary knowledge, but is merely a base for the information they will gain with additional schooling. The class touches on so many ideas, most of which can be examined in much greater detail. I knew from personal experience there was more to cancer than just a mere mutated division of cells, and I recognized how much more there was to know about all topics, so it started to make sense to me that there was more to obesity as well.

I learned that Mr. Gleason was one of three Science Research teachers at the time. Science Research was a class started by students who were interested in pursuing their interest in science in an independent and innovative course. Mr. Yagid helped the founding students create a course that would enable them to study any topic in any scientific field, from stem cells to water purification. Students can choose whatever inspires them, and they research it for three years, beginning in their sophomore year. After completing the second year of the

Science Research program, which is junior year in high school, each student must conduct an experiment so that the following (and final) year they can enter science fairs and compete with their research results.

Each Science Research class has students from all of the three participating grade levels: sophomores, juniors, and seniors. They all support each other much like a self-sustaining community. Presentations to classmates help the students learn professional skills, which are needed later when contact with research scientists becomes necessary. These scientists foster research knowledge by providing access to their published papers with records of their experiments, and by answering questions to further the students' understanding.

When I was a freshman, this class sounded very impressive, but I also considered how much work it would be. I worried that I would not have a topic that I was passionate about enough to research. While I was startled by my realization that I had been raising money for a disease I barely knew anything about, I was not sure I could devote three years to learning about it. Studying fetal development and genetic disorders was also fascinating to me, but I did not think it was interesting enough to dedicate three years to researching. However, I was curious about so many different topics that I hoped I would find a primary interest during my time in the course, *if* I decided to sign up.

It was not until two months later when I left my lunch period to discuss classes for the following year with my guidance counselor that I again considered taking the course. She was new to the school, so I introduced myself and asked her if she had any thoughts about my course load we were designing for the incoming school year.

I sat down in her office and started discussing different ideas. I planned to take Advanced Placement Biology as well as Anatomy and Physiology during junior year, two years away. She questioned my willingness to give up a free period. "I'd be more than happy to sacrifice them," I said honestly. It's not like free periods would help me in the long run anyway.

"You sure love science, huh?" questioned my counselor. "How about we focus on this year for a little while since we only have ten more minutes. Sound good?" She passed me a sheet of paper that listed some of the most popular elective options for incoming sophomores: Culinary, Business, Journalism, Oceanography, Meteorology, Art, and Science Research. I felt like there were only a few that I liked.

Upon reading the last choice, I remembered my Biology class, and what Mr. Gleason and Mr. Yagid had described. I questioned her about the program, and she so intuitively replied: "I do not know exactly what this program entails, but I've heard it's great for people interested in going into science as a career. I really think that you should look into it." Out of curiosity, I picked it for my sophomore year elective. I had been debating it for a while, especially after hearing about it in Mr. Gleason's class. I wondered if it would be as amazing an experience as I had heard it would be from him. Since I had enjoyed Biology, I wanted to take a class where I could pursue my interest and dive into it. Without any further hesitation, I signed up.

Later that month there was a meeting for students new to the program. I introduced myself to the two teachers I had not yet formally met, Mr. Yagid and Mr. Hughes. We were meeting in an eccentrically-decorated Physics classroom in which they taught. The normally boring whiteboards at the front of the classroom were covered in drawings and charts, leaving barely an inch of blandness exposed. The desk in front of these boards had objects stacked up half a foot. They looked like toys to me, but apparently were the materials used for experiments and demonstrations. Posters hung all over the walls and old students' work lay scattered in shelves and on display. The messiness was endlessly entertaining and stimulating.

While Science Research was not held in this classroom, I really started feeling excited about the atmosphere of this part of the science community at my school. I sat down next to the only familiar face, a girl named Sarah.

Students already enrolled in the course, who were completing their first or second year, attended the meeting so they could introduce us to the fundamental ideas of the program. Each gave a basic overview of their research as a general introduction to the kind of work we would be completing in the near future. Students' topics were very complicated, with fourteen-word titles that couldn't help but sound impressive. Most of the new students, my peers, had a list of possible research topics that we rattled off from previous brainstorms. Ideas were just concepts that we found fascinating from Biology or Earth Science classes.

A few weeks later, my mom and I attended a frighteningly well-coordinated "Science Research Symposium," hosted by the current Science Research classes. Students presented their research and discussed articles they had read that were very influential to their research. I was prepared to be impressed with cool experiments and ideas, but I had not expected to be scared by how incredibly knowledgeable the students were about their topics!

Was I going to be in over my head?

One of the students' presentations on the projector failed because of technical difficulties, but she continued without a flinch to describe her complete proposal! I was genuinely astounded by her professional attitude and graduate-level knowledge.

My mom whispered in my ear: "Is this course going to make you like her? Buckle up!" As scared as I was, I was also excited. I wanted to be as comfortable on stage speaking about my own original research as she had been.

I took this fear and turned it into an ambitious energy as I began to search for a research topic. In a few months, I would advance far enough to be able to announce to my classmates that I had chosen my topic for research.

3

NANODIAMONDS

The beginning of my research career

The summer after ninth grade, I began my research by reading news articles about various birth defects and random biology related topics I found interesting. I had no particular interest in one specific topic; I just clicked recently published articles. There were lots of things I found interesting: Down syndrome, autism, fetal development. I have always been curious about birth defects and disabilities, so I used that as my starting point.

As a sophomore, walking into class on the first day, I was incredibly nervous. I looked around the room and saw seniors and juniors who I only recognized from the end-of-year presentations I had watched. Seeing my friend Sarah there made me feel a little more comfortable. Our teacher, Mr. Yagid, had us play a game where we practiced our public speaking skills by sharing about our breakfast with a partner. I'm pretty sure I said something along the lines of "I like waffles." I watched kids talk for minutes about the kinds of foods they like. As hilarious as it was, I was amazed.

Next, older students stated the titles of their research projects to show us all of the possibilities in the world of research. We had freedom to choose anything we wanted.

I expected to hear "Water Purification," but instead I heard things like "The Effect of TiO2/Graphene Oxide on the Purification of Tris(2-Chloroethyl)Phosphate (TCEP) and CuSO4 (CS) Contaminated Water." This motivated me even more to get started researching so I could have a topic title that sounded as impressive as that did.

As the weeks went on, the new students delved into research about each of our general topic ideas. I was researching various birth defects, but I was feeling my passion dwindle as I read these boring articles. I decided to settle on one topic that was more specific so I could read more interesting articles. I chose Down syndrome unofficially. I knew I could make this my topic and thought I could go far with it. I was excited to share this with Mr. Yagid.

In class, Mr. Yagid gave a talk to the new students and explained to us how to find a topic. Students must do a lot of research, and at the same time, they must think long term and plan for their future experiment. I spent that class period contemplating my topic some more. For me, I would not be satisfied with myself in this course if the result of my work did not help anybody. All I wanted was to make a difference. As cheesy as it sounds, that's what my goal was, and still is.

I thought about the logistics of an experiment with Down syndrome. Of course I was nowhere near conducting an experiment, but I had been instructed to think long term. To really research Down syndrome development would take at least a month, because that's how long mice are pregnant for, and those would have to be the test subjects for an experiment of this nature.

I would also have to run tests on the children and the parents. Even if I was able to accomplish that, a "cure" for Down syndrome was never something I was looking for, and I'm not sure is even necessary. This research is ethically controversial, so, ultimately, I chose not to be involved as I understood the debate more. Although this realization made me frustrated because I had to reconsider topic options, it opened doors to another path for me to take.

I had wanted to share my topic with Mr. Yagid at our first biweekly meeting, so I was disappointed that I had to go back to the drawing board. Every two weeks, students meet with the teacher and discuss their progress. The goal of these meetings is to aid the student with a personalized plan for the next two weeks and also to provide the teacher with a progress report. I had really wanted to make a good first impression on Mr. Yagid at our meeting, but now I had to come up with something new. I was so scared that I wouldn't impress my teacher. I did not want to be labeled a slacker for the next three years!

I began scanning science news websites for something interesting. "Tiny diamonds to boost treatment of chemotherapy resistant leukemia" caught my eye. Diamonds! I knew Mr. Yagid would be intrigued. I got to my meeting and I explained what I had done so far this year. I showed him fifteen articles on Down syndrome, four on autism, and one on diamonds used to help treat leukemia.

"Which one's your favorite?" asked Mr. Yagid.

"This one!" I pointed to the one about nanodiamonds and leukemia.

"Why?"

I had not thought about why, I just thought it was an interesting article. "Hmmm…" I thought. It was cool because they used diamonds to treat cancer, but I was more surprised that the treatment of cancer was not well defined enough to know what works and what doesn't.

"I like how even though there are ways to treat almost every type of cancer, they are still searching for ways that could be more effective."

"Perfect! I expect twenty more articles about leukemia treatments by our next biweekly meeting."

The reason I had liked that article became the reason I liked all of the articles I chose next. I was so impressed with all of the new clinical trials and drugs being developed to treat so many different kinds of cancer. Mostly, you hear about radiation and chemotherapy. I had no idea that so much more existed.

The term chemotherapy means chemical treatment, but it has come to be most closely associated with one specific type, cytotoxic chemotherapy for the use of treating cancer.

Cytotoxic chemotherapy works by basically killing all cells that divide rapidly. This is the most common form of cancer treatment, but also can be the most consequential. The drugs I would soon begin studying, Bcl-2 inhibitors and Mcl-1 down-regulators, are pretty much unknown to most people outside the medical community. However, the goal of these drugs is to treat cancer without the harmful side effects.

I know these side effects; I watched my mother lose her hair and feel sick. Every decision one makes is based on their previous experiences, and to me the decision to make my topic cancer drugs just felt right.

4

ADVANCEMENT

In my research

One of the biggest steps in a Science Research student's journey is going from reading lay articles to journal articles. Lay articles, or basic news stories, do not explain the complex experimentation and results that are represented in long, detailed journal articles.

The transition to reading journal articles represents a large step in a student's knowledge of his or her topic: the ability to comprehend a journal article and its vocabulary. Once a student is at this stage, the tasks at hand include contacting researchers in that particular field. Researchers provide articles, answers to questions, and eventually, mentorship for the experiments students conduct.

It took me a week to write my first email. I had a simple request, permission to view one article, but researchers, at least to me, seemed so superior to us students. I asked for samples of requests from other students.

Esther, one of the girls who started the course two years earlier, aided me in this process. It was so helpful to see the way she approached researchers and ultimately how she carried out her own research, which ensured she met the personal goals she set when creating the Science Research course with the teachers.

When I hit send, I felt a rush of nervousness. I was afraid that I would embarrass myself to someone important.

These are the kinds of skills one learns in such an innovative course. From the beginning, Science Research was described to me as a professional, college-level course in which one develops lasting knowledge applicable to the real world. It was moments like this that helped me truly understand that with every email I sent, I was connecting with the future of cancer treatment.

The person I sent that first email to was the head researcher of a large, well-known cancer center. He attends conferences often and presents to other researchers in the field. I had looked into many of these conferences and learned about presenters to understand the most innovative new ideas. Since he was such an important scientist, I wanted to make sure my note was phrased professionally.

I had some seniors in my class edit the email until I thought it was perfect. I got the email approved by Mr. Yagid and I sent it. I had requested an article and within minutes he sent it to me with no note attached to the email. *Oh*, I thought. *That was easy!* So after I finished the article, I asked a question.

The article had been about a drug trial and there were a lot of parts I was confused about. For one, it said the drug had been successful, but then they did not make a big deal about it in the conclusion. To me, it seemed like they were saying the drug being successful at treating the cancer was not a big deal. So I asked the author about it.

The next day, I got his response. It was something along the lines of "We use cultured leukemia cells," and that was all. I felt like his short response meant he was dismissing me for not understanding that they had done the test on cells, not on tumors extracted from patients. Then I realized that all of that information was in the Methodology section of the article that I had skipped because it was too confusing.

I was embarrassed to show Mr. Yagid the response. He did not think it was a big deal, but I did. I never spoke to the researcher again, but I learned from the experience so it was not completely pointless.

Over the next few weeks I read what felt like all of the free articles I could find online about my topic. I then subscribed to Google Scholar Alerts so I could get an update when any articles containing "Bcl-2," "Mcl-1," or "ABT-737" were added to Google Scholar, the part of Google for journal articles and other professional resources. I read the emails on a daily basis, hoping to find more articles to read. At that point, I had to contact more researchers.

The time came when I needed more articles that were not available for free download. I contacted half a dozen researchers, got a handful of articles back and began to read. I saved my questions for the end, so I could ensure that I was not foolishly asking a question before I discovered the answer in my readings.

Upon reaching the end of an article, I tried to look up the answers to my questions, but if they were too hard to find, or the answers were hidden in another article that required a subscription to the journal, I began drafting an email. My email writing skills kept improving and I was learning a lot about the processes and techniques researchers use in their experiments.

As weeks and months passed, I developed a title and abstract for my research. I knew my goals and how to carry them out. I had not designed an exact research plan, but I was getting there.

I kept a handwritten journal and when I found an article I wanted to read, I would make a note of it. If I read an article and liked it, I would write down the authors' names. Now I had folders filled with papers full of lists and articles to read. It was professional and organized. My vocabulary skyrocketed from Biology I Honors to Science Researcher. I kept a list of vocabulary that grew long and contained every word that I would need to read most of the articles I found. I developed as a reader, and also as a listener. I took suggestions from researchers and delved into curiosities about their ideas.

I had grown so much so fast I was even surprising myself. I really felt like I belonged in the class with my ideas and abilities. Not only that, but the sophomore class had begun going out for ice cream together so we could all get to know each other and each other's research a little more. It was a way to connect with the students in the other class and grow as a community. Science Research is a program, not a class, but at times it truly felt like a family. We were a special group of students in our school. Although not everyone in school knew about the course, we knew how exceptional it was.

Of course, there were times when I still felt brainless. The seniors were being interviewed for the town newspaper after school and Mr. Yagid said that day in class, "If any juniors and sophomores want to come, I'm sure you would give an interesting perspective for the article, so the reporter can see the growth from early years to completed research."

That was on Monday, the only day when I had dance for only two hours, and it started later than usual, so I was able to meet with the reporter. "I'll go," I said and so did a few others.

"Great."

After school I walked into the classroom and there were only seniors in there. We sat quietly waiting for the journalist to arrive. After a half hour, he finally arrived. He interviewed the seniors one by one. I did not want to ask to go next because my part was not as important to the article.

I waited for literally hours just listening and when I was ready to leave I said, "Can I quickly describe the early stages of the program and an overview of my research so I can go?" I'm not sure if I sounded annoyed, but I sure felt it. I had dance in an hour and tons of homework to do, but mainly I was frustrated that I had been sitting there waiting so long.

"Alright."

I sat down at his table and started discussing the beginning stages of research and describing how I found my topic. "I already know all about that from an email Mr. Yagid sent me."

"I'll just give you a brief summary of my topic, okay?"

"I guess."

He was not very enthusiastic about my input. I began giving my usual introduction that I give to teachers and my family when they ask. I gave an overview of the need for new cancer treatments and how this treatment would work.

"Alright, finished yet?"

Isn't a journalist supposed to take all of the information they are given?

"Sure." I got up and left. Looking back at the seniors, they seemed about as shocked as I was. The reporter had taken a really long time asking the seniors questions and clarifying the science they were describing, but he did not want to hear about the course or anything I had planned to say.

The next week, we got the newspaper and I was not mentioned in the article. The reporter had focused on the seniors' research projects and that was about it.

I felt very unimportant, but I reassured myself that the man was not a scientist and did not realize the significance of everyone's research. I let that motivate me. I told myself that someday in the future I'd get the opportunity to share my ideas publicly again, so it should not bother me that I spent hours sitting in that classroom, waiting to speak, only to be dismissed.

5

NERVES

For no good reason and for good reason

After researching for months, the sophomores were ready to share what we had learned. It was time to present our research to the class. As about twenty people ran up to the calendar to sign up for a date, I followed the pack. I signed my name on the only open spot I saw, the second day of presentations. I was petrified, because I had only seen the seniors' presentations, and they were always flawless.

The first step to building a presentation is choosing an article to present. I began reading one with a very intriguing title and abstract. As I read, I constructed my presentation. I made it look professional, flow nicely, and have a few strong bullets per slide. I was very proud of the start of my work. Then, as I kept reading the article, I was becoming more and more confused.

The structure of the article was very hard to follow and the experimentation was very unclear. Frustrated because my presentation was in a week, I decided to start over. This time, I decided to focus on the clarity of my presentation. I chose a completely new article. I speed-read it, but then I decided that I had most likely missed a lot of details, so I re-read it many times. I wanted this presentation to be perfect.

That weekend, my family was taking a trip to New York City because my dad was in a bike race. After we sent him off on his race early in the morning, I returned to the hotel while my sisters and mom shopped. I sat cooped up in that hotel room for hours and hours staring at my computer screen.

When my family returned to the hotel room, I practiced my presentation for them. Stammering and stumbling, I jumped from bullet to bullet while my family watched and listened, noticeably confused. For the rest of that week, I stayed up every night working on my presentation.

I woke on the morning of my presentation and jumped out of bed. I dressed up for the occasion and then spent the next twenty minutes going over my presentation again and again. Chemistry class felt like a blur. Waiting outside the Science Research classroom for our teacher, I asked the seniors for advice.

"Talk slow."

"You'll do great, don't worry."

"Relax."

"Just don't mess up!"

While I was standing in front of the class, I was frozen stiff. I knew not many others would be making such a big deal about this because my supportive classmates would never be overly critical about my performance, but I wanted to have a good reputation and this was my chance.

Nervously, I began: "Today I'll be presenting A Novel BH3 Mimetic S1 Potently Induces Bax/Bak-dependent Apoptosis by Targeting Both Bcl-2 and Mcl-1."

Before I knew it, I was finished. "Any questions?" I asked.

"How can the treatment options you are speaking of work better than chemotherapy?" someone asked.

"Typical cytotoxic chemotherapy works by killing all rapidly dividing cells. This proposed Bcl-2 inhibitor is a small molecule drug that works to inhibit the function of specific proteins in cancer cells. Targeting these proteins could kill the cancer cells but would not kill normal cells. This will hopefully limit side effects for people in the future.

"Chemotherapy now is often almost as damaging to a person as the cancer itself," I continued.

"Once these drugs pass their trials, they can be offered as an alternative to cytotoxic chemotherapy. Ideally, these drugs and similar ones will take over the future of cancer treatments."

When the question session finishes, the class evaluates the presentation and gives advice for next time. With little to say other than that I sounded nervous, I was congratulated on a job well done.

That night, Mr. Yagid called my house and told my mom how impressed he was with my presentation. I felt so extremely proud, and was even more motivated to keep working.

The stresses of a presentation are tremendous, but completing one brings major relief. After the relief fades, the next tasks are assigned and the pressure is brought back. My next assignment was entering the Southern Connecticut Invitational Science and Engineering Fair (SCISEF). I began by typing up my introduction and conclusion, but I was stuck on the middle.

For this science fair, new Science Research students propose research that they are planning on conducting later. The research plan is evaluated, not the results, since the project is not completed yet. I decided to wait to write down that portion of my plan, because I was still unsure of what I wanted to do.

For the rest of that week, I designed the layout of my poster, leaving big blank boxes, which stared at me while I thought. As the days flew by, I couldn't decide what I should test. The articles I had been reading were about new drugs that are being developed and their effectiveness. I did not know how to take that further. If they had all been tested, what was left for me to do? Feeling useless, I stalled making my poster.

It was three days before the science fair and I still didn't know what to write. Sarah and I went to Mr. Yagid's office during our free period to discuss our posters with him. She had finished her poster and wanted approval for printing, and I just didn't know what to do next. We entered and sat down. Sarah began, "I finished my poster," and he signaled for her to pass him a copy.

Looking it over briefly, he asked "What makes your research authentic?" imitating a judge at the competition. Frowning, Sarah shrugged and opened up her laptop to make some edits.

Authenticity is a key factor in innovation. To make a difference, you must make a change. Introducing a new factor into the research is what can propel it to success. So when Mr. Yagid asked where my poster was, I knew what to say.

"I've just decided what I want to do. I would like to test the drugs I've read about the most, ABT-737 and Maritoclax. To make this authentic, before the drug tests, I will test survival protein levels in different cancers to see which ones control cell death more in each different type of cancer. If certain cancers have high levels of these proteins, they most likely rely on anti-cell death mechanisms."

"Has that been done before?" he questioned.

"I don't think so. Let me check." I opened up my computer and got to work. Sarah and I spent all of our free period and lunch period in Mr. Yagid's office, typing away. When it was time to go to class, we had both finally created posters we were very proud of.

Over the next two days, most of the students in Science Research printed and mounted their posters. The fair was getting closer and people were becoming more excited. When the day finally arrived, I was dropped off by my mom at school. It was a Saturday morning and it was freezing outside. I saw all of the girls in the program shivering and running into school in high heels and dresses. The bus came to take us to the fair and I boarded with Sarah. For almost the entire bus ride, we sat in nervous silence, running through our presentations in our heads.

Upon arrival at the fair at 9:30 a.m., we signed in and got our nametags and our schedules. Sarah's presentations were scheduled at 10:30 and 11:00. Mine were at 11:30 and 12:00, the two last time slots available.

We walked into the school's gym, where the fair was being held, with the rest of the Science Research students. Mike, a senior, pushed in a cart with all of our binders and posters on it. We all took our things from the cart and set out to find our spots. I was near my friends Nick and Will.

I set up my poster and placed my binder and notebook in front of it. The room was noisy because many students were practicing their presentations aloud (on repeat!) to ensure a smooth performance. I decided to do the same.

A little while later, since neither Sarah nor I were presenting for quite some time, we walked around the gym to see the other competitors' posters. There were so many innovative ideas in one room; it was a lot to take in. From joint implants to the Golden Ratio, we saw almost 200 projects. The proposals and completed projects had the capability to change the world. In that moment, I didn't care whether I placed in the competition or not. The projects were all so impressive, I wanted to talk to every person about theirs!

When 10:30 came, Sarah returned to her poster to present and I watched parts of my friends Esther, John, and Mike's presentations. Then I went over to Sarah's poster to watch her end her presentation. The judges asked her a few questions and she answered them simply.

By the time 11:30 rolled around, I was waiting at my poster, ready to go. The judges approached my poster and introduced themselves. I could feel my hands shaking as I began, "My name is Jennifer Schwartz and today I will be presenting 'Induction of Apoptosis in Leukemias using ABT-737 and Maritoclax.'" From background information to the conclusion, I made sure to expand on every statement so they would know my knowledge of my topic.

At the end, the first judge said, "You understand that one drug will never treat cancer on its own, right? Chemotherapy is delivered as a cocktail of drugs, not just one." I knew ABT-737 was a small molecule inhibitor drug with a specific target, so it has the capability to treat cancer more effectively, even on its own. In that moment, I second guessed myself and I did not speak up. I wondered if maybe it did have to work with other drugs even though I was pretty sure it didn't.

The next judge said, "How will this drug be administered?"

Honestly, I was not sure. I guessed it would be administered like other cancer treatments. "Intravenously."

"Alright, so that's just the same. It would be great if it could be administered subcutaneously, because then I'd be able to provide it at my clinic."

The third judge inched closer to me so he could hear in the very loud room. "How are you measuring the protein levels?"

"Using flow cytometry," I responded with pride that I knew an answer I thought was right.

"With what antibodies?" Again, I didn't know. Filled with embarrassment, I didn't answer.

The judges congratulated me on a job well done and I walked over to Sarah. "How'd it go?" she asked. The judges seemed impressed during my presentation, but then I couldn't answer their questions. I didn't care whether I won, so the question part didn't bother me very much, but it did motivate me to research the answers.

"Alright," was all I could say.

"Same. Wait! Jen, it's 12:00!" Presentations were supposed to be five to ten minutes, but my session took up half an hour. It was already time to present again. With not even enough time to worry, I hurried back to my poster board.

My next presentation took even longer. I had been trying to explain as much as I could and not rush, like I had during my presentation in class. During my second try at the fair, I was able to answer more of the judges' questions. At the end, I was congratulated again and the judges left. One judge stayed back to speak with me. "You know what you should do?" he said. "You should go to a cancer support group and explain your research to them. I'm sure they'd really be proud of all that you've done."

"Thank you, that's a great idea!" I said, but I thought I would be way too scared to present to cancer patients. I know they'd be appreciative of my efforts, but I did not want to seem like I knew things that their doctors did not. Besides, the therapy hasn't been tested on people yet. Also, in a time of suffering and sadness, I thought they would want emotional help, not what I would give them.

After presentations were finished, we packed up our binders and posters and went to the cafeteria for lunch. Everyone in our Science Research program sat down at a table. Sarah, Mike, and I sat toward the end of the table, near some students from another school. We all introduced ourselves and our topics; each topic was equally impressive.

After lunch, we went to the auditorium. A man presented about his experiences in the science world. He had been a TV personality on a science show. He got to witness tons of innovative research projects. If we hadn't already been impressed with the ideas from the day, the videos he showed us proved even more the effects of research on the world. From slime to robots, his projects proved innovation could mean anything.

After his presentation, the award ceremony began. An announcer took the stage. One by one, each school was named and the students from it cheered. Shouts echoed in the large auditorium as tons of people yelled for each school. Ridgefield brought significantly fewer students than most, so we had to fight to make as much noise as our competitors.

However, for the comparatively small number of people who attended, students from our program placed in most of the categories.

"Next up is Proposed Health and Medical, our largest category," said the announcer.

I looked at John and joked, "There are three places, one for each of us." After seeing all of the posters, I knew I had a slim chance of placing, so I did not even hope.

I was proud of our program despite the fact that no sophomores placed. So many upperclassmen were successful that the new students all left feeling inspired. The sophomores were filled with hope.

On our walk out to the bus, Mr. Yagid pulled a few of us aside. "I don't want you guys to be discouraged that you didn't place. I'm very proud of all that you've accomplished in such a short time."

On the bus ride home, Sarah and I could not stop talking about the cool research ideas we saw presented at the fair. When the ride was over, we got out at school and my dad was there waiting to take us both home. He excessively congratulated us on our participation in the fair and our progress in our research. Mr. Yagid waved to my dad and we got in the car.

Sarah and I agreed--we had learned a lot about presenting and even about science in general. We had had a great time despite not placing. The next time we were in class, we both knew we would be inspired to make our research plans even more creative and influential.

6

CHANGES

To my research and thoughts

On Monday, everyone walked into class with a new spirit. The winners were proud and everyone else had a new motivation to excel. I knew I wanted to keep trying hard to make a difference in the science world the way the fair winners had already begun to do.

The next step for me was creating a research plan that would include the new ideas I had been thinking about since the fair took place.

During class that day I read more articles and sent multiple questions to researchers who had performed the studies I had read about.

I had a biweekly meeting later that day and was excited to share my ideas with Mr. Yagid. I went to his office to tell him about the studies that I had just read, and that I wanted to modify my research plan.

This was pretty major, so I knew I wanted his expertise. "Well, your next assignment is a poster for the Symposium," he said, referring to the Science Research program's annual event at our high school showcasing progress made by students over the school year.

"It's due in two weeks. Most people create a poster based on an important article they've read recently. Some sophomores and all juniors make their posters about their research plans. I think you should do a research plan, but you only have two weeks, so if you don't figure out what you want to do by then, then just do your poster on an article."

"Well, I want to do a research plan; I just don't know what to do."

"You can do it. Trust me."

I'd been so caught up in the science fair and my new research plan, I almost forgot about the yearly walk-a-thon that was coming up. I wanted nothing more than to focus on my research rather than go to the walk-a-thon. I was encouraged by research that I could see was actually making changes in the cancer research world. I knew the walk-a-thon event raised millions, but I was really interested in scientific results that I knew would enhance treatments in the future.

While I later learned that a percentage of money raised from the walk-a-thon does go to research, at the time I thought it only went to patient care so I thought my time might be better spent actually doing research. But, since I always have a good time at the walk-a-thon, I decided that it would not hurt to go again. I added it to my calendar and continued with my science work.

I contacted many researchers in the field of cancer medicine and cell death to discuss their most recent studies. I read article after article about new trials that were occurring and new ideas researchers had in order to make sure I would be performing a valuable test when I did my experiment.

I concluded that the drug I wanted to test, ABT-737, was definitely worth investigating. I had not decided exactly how I wanted to experiment, but I knew I wanted to confirm this cancer therapy's validity as a treatment.

By the end of the week I had emailed seven researchers about their work. While waiting for email responses, I texted my friends about the walk-a-thon. We were planning to have a tent with food so we could take a break if we needed one while we were walking. Our team had raised a decent amount of money.

The next couple of weeks flew by. I came so far in my research. From the many articles I had been reading, I discovered that one of the cancer proteins I've been studying, Mcl-1, not only affects cell overgrowth, but also affects the function of a normal cell's mitochondria, an organelle that controls energy production and sometimes death in a cell. That was a major discovery for me because it identified a possible future threat to targeting the mitochondria for cancer treatment.

While reconsidering my research design, I had to make sure there was the smallest margin of error as possible. I could not re-do failed experiments because I would not have the luxury of having access to a lab (apart from the singular time between junior and senior year). There was no way for me to test a theory and move on to something else if necessary. Researchers struggle with failed tests often, then they publish the results and other scientists learn from what occurred.

For me, I was only going to have one opportunity to perform an experiment, so I wanted it to be perfect. I wanted every detail to go exactly as planned in the outline. However unrealistic, that had always been my goal. Each time I found a possible threat to my plan, I made changes so I could still stay close to my original goals. But this time felt different. If there was a possibility Mcl-1 down regulation could cause a person mitochondrial dysfunction, I did not want to be researching a drug that caused that.

I frantically emailed Mr. Yagid about what I had just read. From reading this article, I felt defeated. I felt like it was my own fault for not knowing that there would be side effects of the drugs. Obviously I knew ABT-737 and Maritoclax wouldn't work together to cure a person instantly, but I did have confidence in this treatment. I became so frustrated that I stopped putting in the extra effort of working on my research at home for a while.

In class, I organized my files and binder. I quickly read a couple of articles that had little substance. Bored, I turned to my friend Erin who sat by me. "Are you going on the bus today?" I asked, trying anything to get away from my article. Erin was so fun and talkative, I knew she would start up conversation.

"Nah, not today. I have dive practice right after school."

"Jen, is your poster printed? Then get to work," said Mr. Yagid. Not only was it not printed, it was not exactly started yet....

The bell rang and Mr. Yagid and I walked down to his office. "How's the poster going?"

"Well, I still don't really know what to do yet."

"We'll talk in a minute," he said with a confused look on his face, turning into the Physics classroom he shared with Mr. Hughes. I went down to his office and waited by his door considering how to explain what I'd done in the past two weeks.

The poster was due the next day and I had not yet come up with an idea of what to do. I was so bothered by the fact that the protein Mcl-1 controls normal cell as well as cancer cell function. I assumed that meant it was the end of ABT-737 and Maritoclax as possible treatment options. Every time I thought about it I became frustrated that my work for the past couple of months had been useless. It was May and I had to change my research plan. I felt disappointed.

Mr. Yagid came down the hallway and shrugged. "So what's up with your poster?" He must not have seen my email. He pulled out his keys and unlocked the door. We stepped in and took a seat. I explained again what happened.

"You know I'm not a biology teacher, but if Mcl-1 just helps to control normal cell's mitochondrial function, how do you know mitochondrial dysfunction will be a side effect? Obviously if you take away Mcl-1 and it controls the function, it won't work anymore, but how can you be sure? Is it solely in charge?"

"Well I don't exactly know, but I assume if you take away a protein that aids the process of energy conversion, it won't function exactly the same way. And mitochondrial dysfunction in an entire person is called mitochondrial disease which would be quite a terrible side effect of this treatment, which is exactly what I was trying to combat by researching this drug therapy."

"Assume?"

"Well I guess I don't know."

"That gives you something to research right there."

"To confirm my disbelief in this cancer treatment?" I was frustrated, again.

"Keep researching, I'm sure you'll figure it out. If Mcl-1 was only recently discovered as aiding mitochondrial function, it can't be solely responsible for its downfall as well."

Our discussion had sparked an idea. "What if the promotion of mitochondrial survival was the reason it was accredited to mitochondrial function and energy conversion? That would make sense."

I continued with semi-incomprehensible science babble, "Then, if Mcl-1 was targeted to treat cancer, the healthy cells would probably be fine. They've already done studies down-regulating Mcl-1 to treat cancer, they just haven't measured energy production after. But, if the healthy cells are alive because the treatment didn't affect them, it could not have affected their mitochondria. I could probably confirm this with a test of mitochondrial energy production."

"So you're going to do the same test, but then test the healthy cells' mitochondria after?" Mr. Yagid asked.

I pulled out my computer to write down my ideas quickly. Although I was going to be testing a similar idea, I felt like I was confirming a belief, not a doubt. Either test would have worked for my research, but for me since I only get to experiment one time, I wanted a positive outcome.

"But Jen, it's due tomorrow." I looked up. "I'll give you the weekend. I know most of the sophomores are presenting an article at the Symposium, so if you can do a research plan, I'll give you a little extra time. But I want it completely finished by class on Monday."

"Of course, thanks!"

I used that weekend well. Sitting in my room with articles out and notebooks open, frantically typing, I reminded myself of the weekend I spent in New York like this working on my presentation. At that time, I barely had made a dent in my work or knew what I wanted to do with my research.

In that moment, preparing my poster, I felt committed to my research and to this program. I was a real member and I could explain my topic and what it meant to me, which was something I had not been able to do before. Now it had become a passion. It used to be a "cool new idea in cancer research," but through these journeys I had built a connection with cell death proteins, which, as weird as it was, felt empowering.

I got up from my computer to get my research folder out of my backpack. I pulled it off the hook as I looked at my calendar on the wall. What is that star for? I knew I was so busy in May, I could not remember all of the things I had going on. I checked my phone reminders. 7:00 p.m. walk-a-thon. I ran over to my closet, put on my fundraising team's matching t-shirt and sneakers and I hopped into the car.

When I arrived at the track, my friends came over to talk. We sat in our tent and talked for a long time until the ceremony was about to begin. The ceremony happens as the sun sets, then the track lights up and the walking begins for the evening. My team made its way over to the stage in the center of the field. We sat in the grass and waited for everyone else to arrive. The sun was starting to set. An announcer stepped on stage and in a solemn voice requested silence and respect for everyone who was to speak.

A girl from my high school stepped onto the stage. She told the story, her story, of being diagnosed with cancer at age sixteen. Through surgeries, chemotherapy, pain and suffering, she was able to fight through and reach remission. The crowd was in tears by the end of her speech. Heads were turned down, facing the grass. The large crowd that came out that night was there to support people like her.

In that moment, I knew I was on the right track and that research was what I wanted to do.

Although at this point in the course, I had only been reading articles and writing emails to researchers, it felt like more. I was beginning to understand what it was like to make an impact without having experimented yet. I understood that cancer research changes lives and I wanted to be a part of that.

7

THE SYMPOSIUM

And Karolena

Over those past few weeks, I had become distraught about my topic, but then eager and passionate. It all blurred together into a time of self (and topic) discovery surrounding my Symposium poster.

The Third Annual Science Research Symposium at my school was to be a night of live presentations from the graduating seniors and poster presentations from the other students in the program. It was designed to showcase our research, professionalism, and community of students. Everybody had printed their posters and the event planning committees had the whole night set up to run smoothly. I had designed and printed posters that I hung all over school so students interested in joining the Science Research program could come (and be scared) the way I had.

It was a strange feeling setting up that event. I had changed so much as a student and person in such a short amount of time due to a single year in the program. I'd never been a part of a classroom community in that way before and I had definitely never given so much thought to homework before.

In about nine months I had learned that work is not always about the score you receive. My work ethic was better, as were my presentation skills. After that first presentation I had done another that, like the first, continued to push my confidence all year long. That confidence was shining when I arrived at the Symposium.

On stage, I only had to recite a paragraph I'd memorized, but I had practiced it until it was perfect. I knew my poster like the back of my hand and I knew exactly what to say to people as soon as they stepped up to it. While I was nervous, I was too excited to let it bother me.

I walked into the front entrance of my school with my high heels clicking on the floor. I approached a table with name tags laid out all over. "Barack Obama" read one, because Mr. Yagid had put him on the guest list. This was to remind us to impress because it was not just our parents watching. Also, it would be really cool if the President came to our event.

I threw my script out on the way in. There was no way I would be caught holding it, seeming unprepared. I entered the school auditorium to face all of my Science Research friends, some with their faces in their scripts. Everyone was buzzing with nervous excitement and energy.

Sarah, John, Erin, and I went to the cafeteria, where the poster presentations were to be held. We set up posters and refreshments. I stood in front of mine while Sarah pretended to be an inquisitive parent who would be attending. Most people attending were curious parents who loved asking students about their work. "What is apoptosis?" she asked, trying to warm me up for the evening.

"Apoptosis is programmed cell death. It is a controlled form of dying that a cell must decide and commit to. It is safe for the cells around it, as opposed to necrosis, a type of cell death which isn't. For the purpose of my research, apoptosis is the goal. Cancer cells should die safely, so the patient is damaged as little as possible."

"We're good, don't worry," said Erin. "Let's go greet people at the door."

Following her, I listened as Sarah and John ran through what they were going to say on stage. The role of sophomores in the event was to introduce the graduating class on stage and later speak about our own topics in front of our posters in the cafeteria.

When we returned to the front entrance, some parents were already entering. Most of the parents of the sophomores, who mostly could not drive, had dropped their kids off for set up and waited awhile in their cars. Sarah, John, and I stood by the doors handing out programs while Erin greeted people before they got to us. While two seniors continued to greet attendees, Mr. Yagid and Mr. Hughes pulled aside everyone else. "We know you have all worked so hard on this event and we have no doubt it will run wonderfully," said Mr. Yagid. "If you're unprepared, there's nothing your script can do to help in the next ten minutes, so I would recommend tossing them before guests see."

"I want professionals tonight, but I would not expect anything less from you guys," said Mr. Hughes.

"And with that, let's get started. Good luck!" said Mr. Yagid.

From there, we took the stage. Sitting in the order of people speaking, it felt like a bomb ticking down. First, a student introduced the course and our guest speaker. Our guest speaker gave a fascinating presentation about innovative studies being done and interesting facts about the research world from the perspective of a long-time scientist who had started out as a student researcher, similar to us. After he sat down, another sophomore went up to the microphone.

"Thank you so much for that excellent presentation. Next we're going to hear about this year's graduating seniors. They have been an incredible help this year, especially to the sophomores. These seniors are the first class to go through the course for the full three years, so they were able to help immensely, since they had finished their journey to experimentation. Although their research lives are not ending, we are very sad to see them go. We'd like to personally thank each one of them for all that they have done for this class." He sat back down.

The next person to speak congratulated one of our seniors, Isabelle, on her numerous awards, but they were so extensive, he forgot some of them. He pulled out a piece of paper from his pocket to help him with the rest. For the next senior introduction, the next presenter did the same. Then, Sarah, John and

Erin spoke. They each did very well at not only speaking, but also congratulating each senior they spoke about on their success in the program. Next it was my turn.

"This year I've had the pleasure of getting to know every one of the seniors, including Megan. I know I can speak for all of the sophomores when I say that Megan has been a wonderful guide in this process for each of us. She inspired us from day one with her love of science and her dedication to her topic."

"Aside from the movie *Frozen*," I joked, "Megan's primary focus is on embryonic stem cells because of their pluripotent abilities. She placed third last year in her category at the Southern Connecticut Invitational Science and Engineering Fair and was also a Finalist for Biotechnology at this year's Connecticut Science and Engineering Fair. Megan will be studying at Colgate University this coming fall and I'd like to personally wish her good luck." Stepping down from the podium, I breathed a sigh of relief. My classmates on stage smiled at me. I knew I had done well.

The program was not just helping me present about science. I could now much more easily speak in front of an audience. The nights I spent practicing my presentations had paid off.

After we finished introducing the seniors, two of them, Esther and Devin, played a video slideshow they created to commemorate what Science Research means to them and the senior class. It was complete with short video interviews of the students and pictures of what class is like. It showed us all working and helping each other.

Once the video ended, all of the seniors walked on stage. They presented our teachers with gifts as a thank you for all of their efforts and endless motivation. They announced that they were going to start their presentations in multiple rooms and people could go to whichever they pleased. Sarah and I planned to watch as many as we could by switching in and out of rooms because it was hard to decide who we wanted to watch.

All of the senior presentations were inspiring. They had come so far, and we knew we were on the fast track to the finish line, which they had crossed. When the senior presentations had all finished, all of the Symposium attendees flooded the cafeteria. The sophomores and juniors ran over to their posters so they could be prepared for any visitors.

The first person that came over was Erin's mom. "I see all of those big words, what's your topic?" she asked.

"I'm studying the Bcl-2 family of proteins found in a cell's mitochondria that prevent cells from dying. High levels of these proteins can induce the formation of tumors."

"Oh interesting."

"So what I'm doing is studying a drug combination therapy of ABT-737 and Maritoclax to lower these protein levels, to in turn kill resistant cancer cells with, hopefully, fewer side effects than typical chemotherapy. I'll also be testing one of these possible side effects to confirm it won't be a problem."

"Very impressive, thank you," she said before she moved on to the next poster.

I stood by my poster and waited for the next person to come. People were gathered on the other side of the cafeteria; they just had not made their way over to my side yet. While waiting for them, I went to my friend Kelly's poster. Kelly was in Mr.

Hughes' class, so I did not see her often, but I knew her topic could be very transformative to the medical field. "Hey Kelly!" I said while approaching her poster.

"Hello, do you want me to present?"

"Sure, or at least until people go over to my poster," I said.

"Okay, so I have been researching the applications of cord blood. This treatment could help as therapy for cerebral palsy, stroke, spinal cord injury, and neurodegenerative diseases such as Parkinson's. It is believed that about one in three people could benefit from this kind of medicine. The treatment uses hematopoietic stem cells from the umbilical cord. These cells can differentiate into other cells in the body, including those damaged from these types of diseases."

As she had been presenting to me, a woman and what looked like her daughter had also approached behind me. When Kelly was done speaking, the woman said, "Kelly, great job tonight!"

"Jen, this is my mom and sister, Karolena," she pointed out.

"So you're Jen? I want to go see your presentation, I read all about it in the program."

"Oh thanks so much." Kelly had not yet finished, so she briefly explained the rest of her research plan. Her poster was so filled with information, she knew there was no way she could say it all, so she spoke to me casually, without a script and told me all about the very interesting research she would like to do. While we had been all going out for ice cream together, the two classes were definitely not as familiar with one another as peers in one class are.

I had no idea how impressive her research was and how much she knew about it. She managed to fill a trifold poster board with eighteen point font writing, while most of us used thirty-two! I was astounded and I told her so. Then, her mom asked about my poster.

"It is right over here," I said while walking toward it. We stood in front of my poster and Kelly's mom immediately started talking to me. I was pleased to see someone really get engaged with a student, not just scan the poster board's title and pictures silently while the student rambled on saying complicated words.

"As soon as I saw the words Acute Lymphoblastic Leukemia in your title, I immediately wanted to see what you were doing. You see, my daughter Karolena, Kelly's sister, is in remission from ALL."

"Wow, congratulations!" I said while Karolena turned pink.

"So, before I get started with her story, I'll let you present."

"Alright, so I've been studying a treatment option for various types of leukemia and possibly other kinds of cancers that would likely have fewer side effects than typical cytotoxic chemotherapy. It works by lowering the levels of proteins in the mitochondria that prevent cell death and lead to cancer formation and treatment resistance. By lowering the levels of these proteins, cells can be sensitized and killed."

"That's amazing!" she responded. "During Karolena's treatment, she had a lot of side effects. That was probably the worst part for her. Not only did she lose her hair and white blood cells, but also her esophagus and all around her mouth was all burned by the treatment. It was devastating to watch. Your research gives me hope that someday this suffering will come to an end."

"Wow, thank you. That means so much. Although I know that reading these articles right now is not changing the world, I know that I will be successful if I contribute to the knowledge that could eventually treat leukemia with limited side effects." I spoke from the heart, because I was talking to two people who understood (more than I did) what struggling with cancer was like, and why this research was so important.

It was such a quick conversation, but I felt like it was life changing. To me, Science Research has always been more just a class in school, but I had not really thought a lot about changing the world. Of course, we joke about curing cancer and cleaning out the oceans, but do we believe that we can make an actual change? Finding a single cure has never been my goal; I thought about making a difference in a smaller, more unique way.

From that moment on, my goal became to contribute to and share knowledge about mutation-specific cancer treatments. It was in the days that followed when I reflected on my experience at the Symposium that I understood what a difference it makes when people pursue unique ideas.

The general public's view of cancer is often influenced by the media, and that is far out of my control. But before that night at the Symposium I had no idea how enlightening it could be to share with just two people. That single conversation with Karolena and her mother opened my mind to a new view of life. Never before had I seen pride after such suffering. Since these drugs with fewer side effects have not yet been developed for people to use, Karolena had to suffer through a grueling treatment. Because of this, she pushes research forward with her story. I think Karolena knows that sharing her story, however uncomfortable it may feel to tell, is inspiring.

We see the word "inspiring" all over social media, television, and movies describing people who have suffered from diseases and debilitating conditions. But what I later realized was that mere pictures of people cannot fully inspire the way stories can.

Weeks later, I watched a TED talk, an informative internet video, by a woman with severe physical disabilities. She told her story with this idea in mind. She explained that one singular photo of a person simply makes viewers judge by appearance.

From the point of view of the person with the disability or illness, being an inspiration should not be about appearances, but rather, it should be their words and actions that inspire.

Meeting Karolena and hearing her story inspired me because she showed me specifically how my research would help other people who experienced the same difficulty with cancer treatments as she had. Hearing her describe her painful side effects, just as I had witnessed in my mother years earlier, was a reminder of how important this research is. I was motivated to continue my research in this field because I had learned the impact it would have on others.

I thought back to the time when the science fair judges told me to visit cancer patients so they could hear my story. I never even tried to do it because I thought patients would not be interested in thinking about cancer treatments any longer. I assumed they would be trying to push it out of their memory. But I never realized how beneficial talking to survivors would be to *me*.

8

THE FUTURE

And what I'm going to do with it

Although I have not yet conducted my experiment, I already feel such an immense sense of accomplishment. When I conduct the experiment, no matter the outcome, I will know I have contributed to the eventual development of new cancer medicine. I could never have imagined this just a few years ago. I am not the creator of new drugs, nor am I a doctor, but just to have helped in this lengthy process will be fulfilling.

I'm not sure she knows the true impact she has had on me, but Karolena's story left me with a feeling of hope that is almost impossible for me to describe. After devastation occurs, hope can appear and I was lucky enough that she shared hers with me. Karolena and her mother's appreciation of my work fueled my dedication to this research.

Although research is not a career for everyone, there is so much outside of research that people can do to help cancer patients suffering through tremendously painful treatments. Every article published attracts new readers. Every second patients and survivors spend in support groups can enlighten others in the same situations. This kind of spreading of information furthers the public's understanding of cancer and makes more progress than money does. Just being knowledgeable about cancer tells people suffering with it that you care and want to make a difference.

Through the process of researching cancer as a teen I have learned that science is not just a class, cancer is not just a disease, and inspiration is not just an image. There is so much more of my research journey ahead of me, and I know that there is more to learn and more ways to help change lives. I am sure that I'm heading in the right direction, on the gleaming, hopeful track to innovation and meaningful change.

ABOUT THE AUTHOR

Jennifer lives in Ridgefield, Connecticut with her two sisters and parents. Apart from scientific education, she enjoys dancing on her school's dance team and at a local dance studio. She also enjoys skiing in Vermont with her family on winter weekends. As a very active member in her community, she is involved in multiple volunteer groups such as the Molly Ann Tango Memorial Foundation Junior Board and RHS Gives.

Jennifer's dedication to research has allowed her to take advantage of many amazing opportunities offered to her because of her participation in this program. Jennifer, along with Sarah and John, has started a TED-Ed club at Ridgefield High School, modeled after the famous TED talks, to which they were introduced due to their connection with the innovative Science Research program at Ridgefield High School.